YOUR KNOWLEDGE HAS VALUE

Tashif Ahmad

Damages Assessment of Flood

GRIN Verlag

Bibliografische Information der Deutschen Nationalbibliothek:

Die Deutsche Bibliothek verzeichnet diese Publikation in der Deutschen National-bibliografie; detaillierte bibliografische Daten sind im Internet über http://dnb.d-nb.de/ abrufbar.

Imprint:

Copyright © 2014 GRIN Verlag GmbH
Druck und Bindung: Books on Demand GmbH, Norderstedt Germany
ISBN: 978-3-656-71550-4

This book at GRIN:

http://www.grin.com/en/e-book/278465/damages-assessment-of-flood

GRIN - Your knowledge has value

Der GRIN Verlag publiziert seit 1998 wissenschaftliche Arbeiten von Studenten, Hochschullehrern und anderen Akademikern als eBook und gedrucktes Buch. Die Verlagswebsite www.grin.com ist die ideale Plattform zur Veröffentlichung von Hausarbeiten, Abschlussarbeiten, wissenschaftlichen Aufsätzen, Dissertationen und Fachbüchern.

Damages Assessment of Flood

(A case study of District Chiniot)

Jan, 2014

Department of Geography

Abstract:

Flood damage assessment is the estimation of possible damages which occurred due to the floods of 2010 and 2013. In this study some variable are selected to find out total damages. On the basis of those variables damages can be calculated.

The significance of the research is to calculate the damages in District Chiniot due to the floods in past. The local government will use it as basic information to make policies and implement to overcome the damages of flood. The local government will be able to decrease the damages of flood. The research will also help to aware the people about the loss of the agriculture crops and livestock.

The sole purpose of this research is to protect human being. Because, efforts to predict extreme hazards impacts and damages assessment of subsequent flooding can contribute to more timely and adequate reaction towards prevention and mitigation option. This will be done by apprising information to the concern authorities about the danger caused by the previous floods.

Introduction:

Flood is overflow of water from river due to excessive rainfall and ice melting in the high temperature season. There are three major categories of flood and areal flood is one of them. Areal floods occur when water falls on a flat land which is already saturated. This saturated land prohibits the infiltration of water and flat nature of the land restricts the surface run off. In other words impermeable flat land is suitable for areal flooding. Riverine floods occur when flowing water exceeds the discharge capacity of a stream or a river. Water starts overflowing the leaves of the river and inundates the surrounding areas. Flashflood is another significant type of flood. They occur due to the sudden bursting of the leaves and embankments of the river. Owing to their unpredictable nature Flash floods can be more destructive than riverine floods and areal floods. While riverine and areal floods are predictable but if they are left un assessed they can wreak havoc on a particular area. Therefore their assessment is indispensible to reduce their impacts.

Damage is physical harm that impairs the value, usefulness, or normal function of something. When the capacity of water is increases from the capacity of river it overflows and cause flood in the surrounding areas of river. After this the water covers the surrounding areas and causes a number of damages like damages to life, property, and services. Due to damages people lost their valuable things and are not able to perform any type of activity. In the plane areas water flows everything with it and due to stagnant water it causes many types of problems for local community. The stagnant water causes many types of diseases play havoc with the lives of people andfailure of bridges and cutting of roads unable the commutation of the people from one place to another. In this situation survival of a community is difficult in inundated area. Damage assessment is a method or plan which is used for evaluation of the possible losses caused by flood. Damages assessment of flood is required to find out loses which are caused by flood. With the help of damages assessment plan the local and national government is able to measure damages that a community had to face in the past.

Problem Statement:

Flood is a big concern for the local administration of under developing countries. In these countries mostly rural areas are facing problems of flood. The Chiniot district is located on left bank of the Chenab River. The surrounding villages of Chiniot are severely affected by flood which causes different types of damages. When river Chenaboverflows all areas comes under flood which are in the proximity of the river chanab.Due to flood people face many losses. In the rural areas of district Chiniot most people are performing agricultural activities and tending of a flock. But due to flood they lost their valuable animals like buffalos and the flood destroy their crops and cause epidemic diseases for human, animals and agriculture and due to diseases people loses their lives.Chiniot District comes under the flood whenever river Chanab over flows. Therefor there is need to evaluate possible damages which are faced by the local community. The research is beingconducted for evaluation of damages of flood and will help to overcome the possible damages in future.

Objectives:

- To investigate environmental setup of the study area
- To analyze the damages caused by floods in affected villages
- To determine the total damage caused by flood

Study Area:

Chiniot is a city and administration headquarters of Chiniot District, in the state of Punjab, Pakistan. It is located at 31°43′12″Latitude and 72°58′44″Longitude. Located between the hearts of river Chenab with the heads of small rocky hills, it is known for its wooden furniture architecture which has a great attraction in all over the world.

Chiniot city lies on left bank of the Chenab River inside the small rocky hills, giving a tourism sight and a great attraction for tourists. It is located at the point where Faisalabad-Sargodha road and Lahore-Jhang road meet up each other. It is 158 kilometers in the north-west from Lahore and 38 kilometers in North of Faisalabad. Chiniot city is spread over an area of 10 square kilometers with an average elevation of 179 meters (587 ft.).

District Chiniot has a population of around 1.5 million approximately and has an area of 650891 acres. River Chenab falls 75 k.m in the limits of District Chiniot beginning from Pindi Bhatian to boundary of District Jhang. There are 360 villages/Chaks in this District. The flood in District Chiniot is normal figure of the river Chenab caused mainly by heavy rainfalls in the upper catchment during the moon soon July to September.

Soil:

It contains fertile soil mostly useful for large scale production of wheat, rice. It has become the hub of trade, commerce and education. There are Primary and middle school for boys and girls at government level.

Rainfall:

The rainy season starts in July and ends in September. Annual averageprecipitation in both the district Jhang & Chiniot is 372.3 milimeters (Pakistan Meteorological Department Karachi 1961-90). More rains occur in July and August than any other months. Most of the winter rains are received in the months of January, February and March.

Vegetation cover and trees:

The district's trees consist of jand (Prosopis spicigera), karir (Capparis aphylla), beri (Zizyphus jujuba), van (Salvadora oleoides), kikar (Acacia nilotica), shisham (Dalbergia sissoo) and aak (Calotropois spp). Various herbs can also be found, including harmal, akrey and bathoo.

Map of District Chiniot:

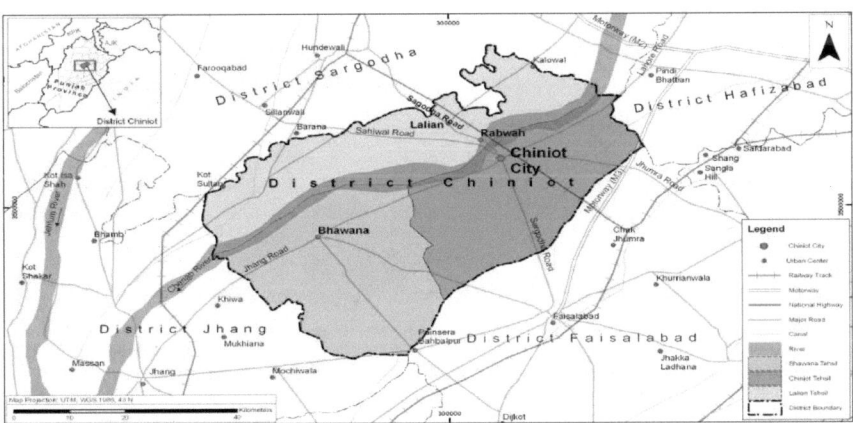

Figure 1.1 Map of district chiniot

Source: http://en.wikipedia.org/wiki/Chiniot

Literature review:

The population places and the liability of human lives have grown severely over the previous half century. The growths of calamity losses are a huge part due to human decisions and reserves. Disaster is different in scale among very huge and smaller. The Venezuela floods in December 1999 estimate 30,000 lives damaged 23,200 houses and demolished an extra 64,700. Minor but still devastating 2003 in Bohorok or Indonesia around 170 people were registered dead later than flash flood. Or 450 homes other structure was damaged beside with 35 alternative cottages, eight bridges, and two mosques. When a wall of water is strike in North Sumatral town. Irrespective of dimension it is need for corresponding efforts in capable and effected counteract evaluate collection based on the assessment of pre-and post-event strategies. The result is impact of a disaster dependence on the country's economic power earlier to the disaster. (Hansson.k et al 2008)

The floods in the Chenab River submerged 100 villages in Chiniot displacing 11,985 people and completely washing away 76 houses and crops on 18,806 acres. The district administration estimated that Rs12.1 million would be needed for houses reconstruction and Rs37.6 million for crop compensation. However, the affected people say their losses were underestimated as not only they lost their ripe crops but also could not cultivate the rabbi crops since the floodwaters receded very slowly. The calamity also destroyed over a dozen of school buildings and roads which have yet to be repaired. (Dawn News, 2011)

In 2010 Flood Pakistan starts in July resulting serious monsoon rains in KPK and Baluchistan. Present estimation illustrate that about 2000, people have died and above million homes smashed. (Moledina, 2010)

Methodology:

Selection of area:

The step first in the research is selection of the study area weather that particular is facing such kind of problem which we are going to research. In any type of research project selection of specific study area has great importance.

Selection of Variables:

In this research some variable will select for the calculation of damages which are mentioned above.

Methods of Data Collection:

Primary Data:

The primary data is first hand which is collected through different sources. For primary data a questionnaire can be generated for acquiring suggestions of community about particular disaster or interviews can conducted of different district officers.

Secondary data:

In the secondary data, data collects from different organizations or departments. In the secondary flood damages assessment reports can be collected or book, articles and research projects can also be collected for secondary data

Spatial Data Collection:

In the spatial data collection satellite images can download. These satellite images can be downloaded from Google earth pro with the help of stich map.

Geo-referencing:

After the downloading these images can geo-referenced. For geo-referencing images are added in the ArcGIS and a co-ordinate system is assigned to the image and then four control points are assigned to the image then this image will geo-reference.

Digitization:

After geo-referencing this image can digitize. In the digitization different shape files can be generated. These shape files are generated on arc catalog and then are added into arc map and then one by one all features are digitized on the image and then all edits are saved.

Damages of Flood 2007:

In the flood twenty four villages were completely damage and twenty four villages were partially damaged. The total area which was damaged in flood it was 46958. And the population was affected it was 98000.

No of Village Affected Completely	No of Village Affected Partially	Total Area Affected	Total Population Affected
24	24	46958	98000

Source: National Disaster Management Authority

http://www.ndma.gov.pk/

Cattle Affected:

A large number of cattle were affected by flood. In which the number of buffs was 30430.The 16867 cows were affected in flood .The total number of sheep was 6002 and the total number of goats were 9400. The total number of poultry was 17241.

Buffs	Cows	Sheep	Goat	Poultry
30430	16867	6002	9400	17241

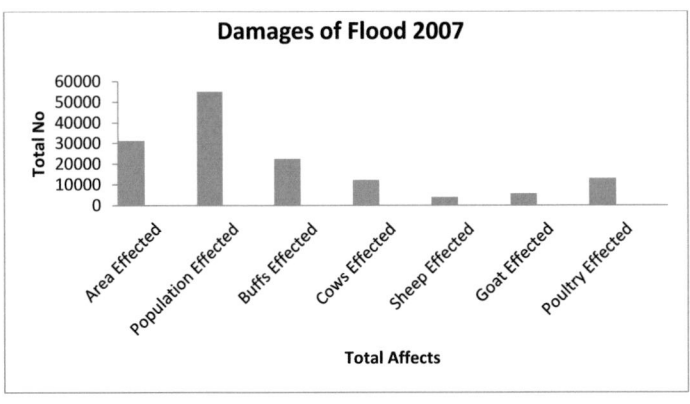

Source: National Disaster Management Authority

http://www.ndma.gov.pk/

Damages of flood 2013:

Mouzas Affected:

Flood caused different types of damages. In the flood nine mouza were totally affected. This 3% of the total mouza. And due to flood in the 10 mouza where houses were damaged.

No of Mouza Affected	9
Percentage Mouzas Affected	3%
No. of Mouza Where House Damaged	10

House Damaged:

In different villages a large number of people were affected. The nineteen people whose houses were damaged and this 76% of total damage which caused in different villages. In the flood five kacha houses were partially damaged and thirteen kacha houses which were totally damaged. The no of pacca house was only one which was damaged in flood

No. of Person Whose House Damage	19
Percentage of Damaged Houses	76%
Kacha House Partially Damage	5
Kacha House Totally Damage	13
Pacca House Partial Damage	1

Total Land and Crop Affected:

The flood caused on large area and damaged a large number of crops. The total land of owner farmer (Acres) 38599 which was under the flood. The large numbers of crops 1041 acres were damaged in flood.

Total Land of owner/Farmer	38599
Crop Affected Area	1041

Person Affected:

Due to flood mostly people were migrated toward safe area. In the flood 411 person were damaged.

No.of Person Affected	411

Total Damage of Flood 2013:

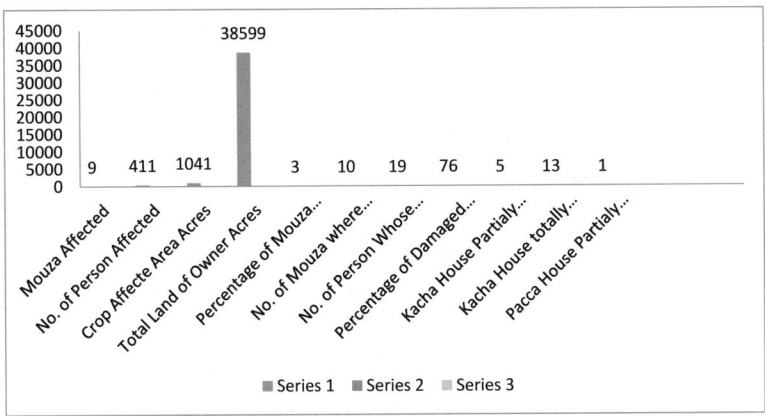

Source: National Disaster Management Authority

http://www.ndma.gov.pk/

Conclusion:

Due to flood people were faced large number of losses. They lost their valuable things in the Flood. Mostly People were migrated toward different area on temporally basis. The stagnant water was caused a number of diseases. The main reason of flood was heavy rain fall and over flow of river.

References

- K.Hansson, Danielson and Ekanberg, A framework for evaluation of flood management strategies, journal of environment management, 8 Dec 2006, pp. 465-480, wwwelsevier.com/locate/jenvman

- BlongRussel& Gissing Andrew (2010) "accounting for variability in commercial flood estimation" Australian Geographer 35:2, 209-222

- De Leo Rob, Daunta, Associated benefits and costs of the Canadian flood damage reduction program , 28 Aril 2000,www.elsevier.com/locate/apgeog

- Dr Ahmad Nazir, "Water Resource of Pakistan" Mirajuddin press, Lahore September 1993.

- StanelyA.Changnon, Jeohowe, lain which (1996). The grate flood of 1993: causes, impacts and Response

- http://www.ndma.gov.pk/

  (National Disaster Management Authority Pakistan)

- Moledian, s (2010). Pakistan Selab (Flood)

- http://www.dawn.com/news/601000/world-forgets-chiniot-flood-victims

- http://en.wikipedia.org/wiki/Chiniot